Saja Waleed Naif Mohammed

Study of the Hpre-Open Sets in Topological Spaces

Saja Waleed Naif Mohammed

Study of the Hpre-Open Sets in Topological Spaces

Definitions, Properties, and Fundamental Results

Noor Publishing

Cover image: www.ingimage.com

Publisher:
Noor Publishing
is a trademark of
Dodo Books Indian Ocean Ltd. and OmniScriptum S.R.L publishing group

120 High Road, East Finchley, London, N2 9ED, United Kingdom
Str. Armeneasca 28/1, office 1, Chisinau MD-2012, Republic of Moldova, Europe
Managing Directors: Ieva Konstantinova, Victoria Ursu
info@omniscriptum.com

Printed at: see last page
ISBN: 978-620-9-53195-8

Study of the $Hpre$-Open Sets inTopological Spaces

By

Saja Waleed Naif Mohammed

List of Symbols

symbol	Description
$\boldsymbol{preC(X)}$	denoted the family of all *pre*- closedsets
$\boldsymbol{preO(X)}$	denoted the family of all *pre*-open sets
$\boldsymbol{GC(X)}$	denotes the family of all *g*-closed sets
$\boldsymbol{GO(X)}$	denotes the family of all *g*-open sets
$\boldsymbol{G(f)}$	graph
$\boldsymbol{HpreC(X)}$	the family of all *Hpre*-closed sets
$\boldsymbol{HpreO(X)}$	denote the collection of all *Hpre*-open subset in X
$\boldsymbol{Cl_{pre}(A)}$	Is the intersection of all *pre*-closed sets containing A
$\boldsymbol{Int_{pre}(A)}$	Is the union of all *pre*-open sets contained in A
$\boldsymbol{Cl^H_{pre}(A)}$	The *Hpre*-closure of A
$\boldsymbol{Int^H_{pre}(A)}$	The *Hpre*-interior of A
$\boldsymbol{N_{Hpre}(X)}$	Be the collection of all *Hpre*-neighbourhood
$\boldsymbol{T_i}$**-spaces**	Respect to *Hpre*-open sets ,for i=1,2

List of Contents

Abstract

The main concept in our work to connect between the definitions of -open sets and - closed sets in the topological spaces with a new definition is called -open set . This new definition was used to define the properties of closure and interior of the sets and also to introduce a new concept the continuous, namely -continuous function, - open function and - closed function and have proved some theorems about these notion .

Also, we introduced the idea of the separation axioms by - open sets , called *Hpre*- T_1 spaces and *Hpre* - T_2 spaces and studied some of their applications .

Finally , we present the -closed graph and strongly *Hpre*-closed graph by - *Hpre* closed sets and we have some results and relationships .

Introduction

The generalized closed sets in topological spaces played a great rol in all topics of general topology in which continuity separation axioms , graph function , connected , etc.

The first to introduce the definition of generalized closed sets ***g*** **-closed** by N. Levine [11] in 1970.

L. Herrington and P. Long [9] in 1975 defined -closed set by using the***g*** -closed set defined by N. Levine in [11] and gave the concept of strongly closed graph.

Dunham and N. Levine [6] in 1980 gave a new results about the generalized closed sets, and studied some of the immovable of this new topology. Also Dunham [5] in 1982 used the generalized closed sets to define a new closure operator , to obtain on the new topological spaces.

In 1982 A.S Mashour, M.E.A. El-Monsefand S.N El-Deeb [15] introduced the nation of *pre*-**open** set and used it to define new types of the continuity which are called - continuous function , *pre* -open function and *pre* -closed function in the topological space.

In 1986 D. Andruevic [1] used ideas *pre* -open set , to define new types of the sets are named semi *pre* - open sets and used they to offer a new kind of the continuity.

Kar and P. Bhattacharyya [10] in 1990 continued the work of A.S. mashour , M.E.A. El-Monsef and S.N. El-Deeb [15] on the pre -open sets and defined another set of separation axioms similar to the separation axioms defined by S.N. Mashhour and R. Prasad introduced a type of separation axioms are called $pre - T_1$ and $pre - T_2$ using the pre -open sets defined by A.S .Mashour M.E.A. El-Mounsef and S.N. El-Deeb [15] .

In 1996 H. Maki , J- Umehara and T. Noiri [13] introduced the spaces.

In 2005 N. Bandyopadhyay and P. Bahttacharyya [3] defined the pre -closed graph and proved the relationships between such sets and the pre -open function.

P. Padma , S. Jackson rani , A. Senbagapriya and B. Ajitha [20] developed the properties of the sets of the pre- open set and g-closed set.

This work deals with some study of the basic concepts in the topological space by using -open set.

Throughout our work , and stands for a topological space (briefly and) and no separation axioms . For a subset of the closure of and the interior of are denoted by and respectively.

In chapter one we introduced a definition of the $Hpre$-closed set and we prove some results , properties and we study a relationship with closed and between $Hpre$ -closed set and -closed set . Next , we define the –open (neighborhood , closure and interior) and gave its properties also give some examples to illustrate some cases . Once again we introduced the -continuty (open and closed) function and we study it properties and the relationship with$Hpre$ -open ,$Hpre$ -closed and $Hpre$-closed functions.

In chapter two we use the open set to define T_1-space andT_2- space and show some theorems ,examples and we show the diagram between$pre - T_1$ -space and$Hpre - T_1$ -space and$g - T_1$ -space this diagram is true but the converse is not true. After that we define the$Hpre$ -closed graph and present some theorems and examples. Also we define strongly-closed and present some theorems and properties.

Chapter One

An Overview of *Hpre*-Open Set in Topological Space

1.1. Introduction

In order to highlight the $Hpre$-closed set and prepare the background for our work we understand the need to recall some definitions , results and prove some theorems related to the basic concepts of our work .

1.2. Forms of the pre-Open and Generalized Closed Sets

Let us introduce these notions in the following definitions :

Definition 1.2.1. Let (X, τ) be a topological space and $A \subseteq X$. Then:

(1) A is pre-open [15] set if $A \subseteq Int(Cl(A))$. And pre-closed if $Cl(Int(A)) \subseteq A$.

(2) A is generalized closed [11] (briefly g-closed) if $Cl(A) \subseteq U$ whenever $A \subseteq U$ and U is open set in X. And A is generalized open ((briefly g-open) if $X \setminus A$ is g-closed.

The pre-interior of a subset A of X [15] is the union of all pre-open sets contained in A and is denoted by $Int_{pre}(A)$ and the pre-closure of a subset A of X [15] is the intersection of all pre-closed sets containing A and is denoted by $Cl_{pre}(A)$. The g-closure of A in X [5] is the intersection of all g-closed sets containing A and is denoted by $Cl_g(A)$ and the union of all g-

open sets contain in A [5] is called g-interior and is denoted by $Int_g(A)$. Now we need the following notation:

Notation 1.2.2. Let (X, τ)be a topological space.

- $preC(X)$ (resp. $preO(X)$) denotes the family of all *pre*-closed sets (resp. *pre*-open sets).
- $GC(X)$ (resp. $GO(X)$) denotes the family of all g-closed sets (resp. g-open sets).
- $preO(X,x) = \{A \in preO(X) : x \in A\}$, $O(X,x) = \{A \in \tau : x \in A\}$ and $preC(X,x) = \{A \in preC(X) : x \in A\}$.
- $GO(X,x) = \{A \in GO(X) : x \in A\}andGC(X,x) = \{A \in GC(X) : x \in A\}$.
- $HpreO(X,x) = \{A \in HpreO(X) : x \in A\}andHpreC(X,x) = \{A \in HpreC(X) : x \in A\}$.

Definition 1.2.3. A function $f: X \to Y$ is called:

(1)*pre*-continuous [15] (resp. g-continuous [2]) if $f^{-1}(U)$ is *pre*-open (resp. g-open) in X for each open subset Uin Y.

(2)*pre*-open [15] (resp. *pre*-closed [15]) if $f(U)$ is *pre*-open (resp. *pre*-closed) in Y for open open (resp. closed) subset Uin X.

(3) g-open [14] (resp. g-closed [14]) if $f(U)$is g-open (resp. g-closed) in Y for each open (resp. closed) subset U in X.

Definition 1.2.4. A topological space (X, τ) is called:

(1)pre-T_1 [19,10] (resp. g-T_1 [4]) if for each $x \neq y \in X$, there exist a pre-open (resp. g-open) set U in X, such that $x \in U, y \notin U$ and there exsit a pre-open (resp. g-open) set V in X, such that $y \in V, x \notin V$.

(2)pre-T_2 [19,10] (resp. g-T_2 [4]) if for each $x \neq y \in X$, there exist a pre-open (resp. g-open) sets U and V such that $x \in U, y \in V$ and $U \cap V = \emptyset$.

Remark 1.2.5. Let (X, τ) be a topological space then :

(1) Any T_1- space [15,11] is pre-T_1 space.

(2) Any T_2- space [15,11] is pre-T_2 space.

Remark 1.2.6.

(1) Any open set is pre-open set [15]

(2) Any open set is g-open set.[14]

(3) Any closed set is g-closed [5]

Lemma 1.2.7. [7] Let (X, τ) be a topological space . If G is an open set ,$CL(G \cap A) = CL(G \cap CL(A))$ and hence $G \cap CL(A) \subseteq CL(G \cap A)$ for any subset A .

Definition 1.2.8. A function $f : X \to Y$ we call that :

(1) The subset $\{(x, f(x)): x \in X\}$ of the product space $X \times Y$ is called the graph of [12] and denoted by $G(f)$.

(2) A closed graph [12] if its graph $G(f)$ is closed sets in the product space $X \times Y$.

(3) A strongly closed graph [9] if for each point $(x, y) \in G(f)$, there exists open sets $A \subset X$ and $B \subset Y$ containing x and y , respectively , such that $A \times Cl(B)) \cap G(f) = \emptyset$.

Lemma 1.2.9. A function $f : X \to Y$ has a closed graph [12] if for each $(x, y) \in (X \times Y)\backslash G(f)$, there exist $A \in O(X, x)$ and $B \in O(Y, y)$ such that $f(A) \cap B = \emptyset$.

Lemma 1.2.10. The graph $G(f)$ is strongly closed [18] if and only if for each point $(x, y) \notin G(f)$, there exist open sets $A \subseteq X$ and $B \subseteq Y$ containing x and y , respectively , such that $f(A) \cap Cl(B) = \emptyset$.

Definition 1.2.11. A function $f : X \to Y$ is called strongly *pre*-closed graph [19] if for each $(x, y) \in (X, y)\backslash G(f)$, there exist $A \in preO(X, x)$ and $B \in O(X, x)$ such that $\left(A \times Cl(B)\right) \cap G(f) = \emptyset$.

1.3. *Hpre*-Closed Sets

In this section we define *Hpre*-closed sets also introduce some properties, examples and theorems.

Definition 1.3.1.[17] A subset A of a space X is called *Hpre*-closed set if $Cl_g\left(Int(A)\right) \subseteq A$.

The family of all *Hpre*-closed sets is denoted by $HpreC(X)$.

Lemma 1.3.2. If B is an g-closed subset of$X$ such that $Cl_g(Int(B)) \subseteq A \subseteq B$,then A is $Hpre$-closed set.

Proof.

Since B is$g$-closed, $Cl_g(B) = B.$ Then $Cl_g(Int(A)) \subseteq Cl_g(Int(B)) \subseteq A,$ so A is $Hpre$-closed set.

Remark 1.3.3. The converse of Lemma (1.3.2) is not true as shown by the following example.

Example 1.3.4. Let $X = \{x_1, x_2, x_3\}$and $\tau = \{X, \phi, \{x_1\}, \{x_1, x_2\}\}$. Then

$\tau^c = \{X, \phi, \{x_3\}, \{x_2, x_3\}\}, preC(X) = \{X, \emptyset, \{x_2\}, \{x_3\}, \{x_2, x_3\}\},$

$preO(X) = \{X, \emptyset, \{x_1\}, \{x_1, x_2\}, \{x_1, x_3\}\},$

$GC(X) = \{X, \phi, \{x_3\}, \{x_1, x_3\}, \{x_2, x_3\}\},$

$GO(X) = \{X, \emptyset, \{x_1\}, \{x_2\}, \{x_1, x_2\}\},$

$HpreC(X) = \{X, \emptyset, \{x_2\}, \{x_3\}, \{x_1, x_3\}, \{x_2, x_3\}\},$ $HpreO(X) = \{X, \emptyset, \{x_1\}, \{x_2\}, \{x_1, x_2\}, \{x_1, x_3\}\}$. So $\{x_3\} \in HpreC(X)$and $\{x_1, x_3\} \in GC(X),$ but $Cl_g(Int\{x_1, x_3\}) = \{x_1, x_3\} \not\subset \{x_3\} \subset \{x_1, x_3\}$.

Theorem 1.3.5. Let (X, τ) be a topological space, then

(1) Any pre-closed subset of X is $Hpre$-closed set.

(2) Any g-closed subset of X is $Hpre$-closed set.

Proof.

(1) Suppose that A is pre-closed subset of X, since any closed set is g-closed, $Cl_g(A) \subseteq Cl(A)$[5 , Lemma 3.3], then $Cl(Int(A)) \subseteq A$, this implies $Cl_g(Int(A)) \subseteq Cl(Int(A)) \subseteq A$. Then A is $Hpre$-closed set.

(2) Since $Int(A) \subseteq A$ and suppose that A is g-closed then $Cl_g(A) = A$ [5, Theorem 3.7], so $Cl_g(Int(A)) \subseteq Cl_g(A) = A$. Then A is $Hpre$-closed set.

Remark 1.3.6. The opposite of Theorem 1.3.5 is nottrue. In Example (1.3.4) note that, $\{x_1, x_3\} \in HpreC(X)$, but $\{x_1, x_3\} \notin preC(X)$ and $\{x_2\} \in HpreC(X)$, but $\{x_2\} \notin GC(X)$.

The above outcomes can be outlined in the accompanying chart, however the opposite is nottrue.

$$pre\text{-closed set} \rightarrow Hpre\text{-closed set} \longleftarrow g\text{-closed set}$$

Theorem 1.3.7. If $\{G_i : i \in \Delta\}$ is a collection of $Hpre$-closed sets of a space (X, τ), then $\bigcap_{i \in \Delta} G_i$ is $Hpre$-closed set.

Proof.

Since $Cl_g(Int(G_i)) \subseteq G_i$ for every $i \in \Delta$, and $\bigcap_{i \in \Delta} G_i \subseteq G_i$, for every i, $Int(\bigcap_{i \in \Delta} G_i) \subseteq Int(G_i)$. So $Int(\bigcap_{i \in \Delta} G_i) \subseteq \bigcap_{i \in \Delta} Int(G_i)$. Hence$Cl_g(Int(\bigcap_{i \in \Delta} G_i)) \subseteq Cl_g(\bigcap_{i \in \Delta} Int(G_i)) \subseteq \bigcap_{i \in \Delta} Cl_g(Int(G_i)) \subseteq \bigcap_{i \in \Delta} G_i$, therefore $\bigcap_{i \in \Delta} G_i$is $Hpre$-closed set.

Remark 1.3.8. The union of two $Hpre$-closed sets it is not necessary be $Hpre$-closed set. As can be seen in the following example.

Example 1.3.9. Let $X = \{x_1, x_2, x_3\}$and $\tau = \{X, \phi, \{x_2, x_3\}\}$. Then $\tau^c = \{X, \phi, \{x_1\}\}$, $preC(X) = HpreC(X) = \{X, \emptyset, \{x_1\}, \{x_2\},$
$\{x_3\}, \{x_1, x_2\}, \{x_1, x_3\}\}$,
$preO(X) = HpreO(X) = \{X, \emptyset, \{x_2\}, \{x_3\}, \{x_1, x_2\}, \{x_1, x_3\},$
$\{x_2, x_3\}\}$,$GC(X) = \{X, \phi, \{x_1\}, \{x_1, x_2\}, \{x_1, x_3\}\}$, $GO(X) =$
$\{X, \emptyset, \{x_2\}, \{x_3\}, \{x_2, x_3\}\}$. Where$\{x_2\}$ and $\{x_3\}$ are $Hpre$-closed sets, but $\{x_2\} \cup \{x_3\} = \{x_2, x_3\}$ is not $Hpre$-closed set.

Corollary 1.3.10. The intersection of an $Hpre$-closed set and a pre-closed set is $Hpre$-closed set.

Proof.

Direct from Theorem 1.3.5(1) and Theorem 1.3.7.

Corollary 1.3.11. The intersection of an $Hpre$-closed set and a g-closed set is $Hpre$-closed set.

Proof.

Directly from Theorem 1.3.5(2) and Theorem1.3.7.

Definition 1.3.12. A subset A of a space X. The $Hpre$-closure of A is denoted by $Cl^H_{pre}(A)$ is the intersection of all $Hpre$-closed sets in Xcontaining A. That is $Cl^H_{pre}(A) = \cap \{G : A \subseteq G, G \in HpreC(X)\}$.

Theorem 1.3.13. A subset $A$of a spaces$X$. Then A is $Hpre$-closed set in X if and only if $Cl_{pre}^{H}(A) = A$.

Proof.

Let A be$Hpre$-closed set. By Definition 1.3.12 $Cl_{pre}^{H}(A) = A$. Conversely, suppose $Cl_{pre}^{H}(A) = A$. By Theorem 3.7 then A is $Hpre$-closed set.

Theorem 1.3.14. let$A$ and B be a subsets of X . Then the following results hold.

(1) $A \subseteq Cl_{pre}^{H}(A) \subseteq Cl_{pre}(A)$ and $Cl_{pre}^{H}(A) \subseteq Cl_{g}(A)$.

(2)$Cl_{pre}^{H}(\emptyset) = \emptyset$ -and $Cl_{pre}^{H}(X) = X$.

(3)$Cl_{pre}^{H}\left(Cl_{pre}^{H}(A)\right) = Cl_{pre}^{H}(A)$.

(4)$If\ A \subseteq B\ , then\ Cl_{pre}^{H}(A) \subseteq Cl_{pre}^{H}(B)$.

(5)$Cl_{pre}^{H}(A) \cup Cl_{pre}^{H}(B) \subseteq Cl_{pre}^{H}(A \cup B)$.

(6)$Cl_{pre}^{H}(A \cap B) \subseteq Cl_{pre}^{H}(A) \cap Cl_{pre}^{H}(B)$.

Proof .

(1) Directly from Theorem 1.3.5 (1) and (2) , respectively. Note that in general $Cl_{pre}^{H}(A) \nsubseteq$ A .For Example consider in (1.3.4) ,let $A = \{x_1\}$, then$Cl_{pre}^{H}(A)= \{x_1,x_3\}$, so $A \subseteq Cl_{pre}^{H}(A)$ but $Cl_{pre}^{H}(A) \nsubseteq A$.

(2) Since $\emptyset$ and X are $Hpre$- closed sets , then by Theorem(1.3.13) (2) is satisfied.

(3) Since $Cl^H_{pre}(A)$ is $Hpre$-closed set , there for by Theorem (1.3.13), we have Cl^H_{pre} ($Cl^H_{pre}(A)$)= $Cl^H_{pre}(A)$.

(4) Since $A \subseteq B \subseteq Cl^H_{pre}(B)$, then $A \subseteq Cl^H_{pre}(B)$. Therefore by (3) , so$Cl^H_{pre}(A) \subseteq Cl^H_{pre}(Cl^H_{pre}(B)) = Cl^H_{pre}(B)$. Note that the opposite direction of (4) is nottrue . For Example (1.3.4)

$A \nsubseteq B$,then$Cl^H_{pre}(A)$⊆$Cl^H_{pre}(B)$, if $A = \{x_3\}$ and $B = \{x_1\}$, $Cl^H_{pre}(A)$=$\{x_3\}$ and $Cl^H_{pre}(B)$=$\{x_1,x_3\}$.

(5)and **(6)** Directly from (4).

Remark 1.3.15.

The equality in the statement (5) need not be true as seen from Example (1.3.9), let A =$\{x_2\}$, B =$\{x_3\}$ and $A \cup B$ =$\{x_2 ,x_3\}$, $Cl^H_{pre}(A)$=$\{x_2\}$, $Cl^H_{pre}(B)$=$\{x_3\}$ and $Cl^H_{pre}(A \cup B)$= X, $Cl^H_{pre}(A)$∪$Cl^H_{pre}(B)$=$\{x_2, x_3\}$, therefore $Cl^H_{pre}(A) \cup Cl^H_{pre}(B) \neq Cl^H_{pre}(A \cup B)$

Remark 1.3.16

.The equality in the statement (6) need not be true as seen from Example (1.3.4) , let A =$\{x_3\}$ and $B = \{x_1\}$ so $A \cap B = \emptyset$, $Cl^H_{pre}(A \cap B) = \emptyset$, $Cl^H_{pre}(A) = \{x_3\}$, $Cl^H_{pre}(B) = \{x_1, x_3\}$,

$Cl^H_{pre}(A)$∩ $Cl^H_{pre}(B)$ $= \{x_3\}$, therefore $Cl^H_{pre}(A)$∩ $Cl^H_{pre}(B)$ $\neq Cl^H_{pre}(A \cap B)$

1.4. *Hpre*-Open Sets

In this section we will study *Hpre*-open sets, review the basic properties and some important theorems.

Definition 1.4.1.A subset A of a space X is called Hpre-open if X \ A is Hpre-closed. We denote the collection of all an Hpre-opensubsets in X is $HpreO(X)$

Lemma 1.4.2. A subset A of a space X, then

(1) $Cl_g A^c = [Int_g A]^c$

(2)$Int_g A^c = [Cl_g A]^c$.

(3)$X \setminus Cl_g(X \setminus A) = Int_g(A)$.

(4)$X \setminus Int_g(X \setminus A) = Cl_g(A)$.

Proof.

(1) Let x is an g-interior of A^c is g-open set of x containing no point of , that is $x \in Cl_g(A^c)$. Conversely , suppose that x is not g-interior of A. Then there exists an g-open G of x which contains no point of . This implies $\in N \subset A^c$. It follows that A^c is g-open of x and consequently x is an g-interior point of A^c that is , $Int_g(A) = (Cl_g(A^c))^c$, this means $Cl_g(A^c) = [Int_g(A)]^c$

(2)let$x \in (Cl_g(A))^c$then $x \notin Cl_g(A)$ since there exist g-open U , $x \in U$ such that $x \in U \subseteq A^c$ this implies $x \in Int_g A^c$ therefore $(Cl_g(A))^c = Int_g(A^c)$.

(3)A same proof of 1.

(4)A same proof of 2.

Corollary 1.4.3. Let (X,τ) be a topological space . Then the intersection of an $Hpre$-open set and an open set is $Hpre$-open set .

Proof.

Let A be a $Hpre$-open set and G be an open set in .then $A \subseteq Int_g(Cl(A))$ and $Int_g(G) = G$, by Lemma 1.2.6 , we obtain that $G \cap A \subseteq Int_g(G) \cap Int_g(Cl(A)) \subseteq Int_g(G \cap Cl(A) \subseteq Int_g(Cl(G \cap A))$. So that , $G \cap A$ is $Hpre$-open set .

Theorem 1.4.4. A subset A of a space X is $Hpre$-open if and only if $A \subseteq Int_g(Cl(A))$.

Proof.

Let A be Hpre-open subset of X. Then $X \setminus A$ is Hpre-closed and $Cl_g(Int(X \setminus A)) \subseteq X \setminus A$. By Lemma 1.4.2 $A \subseteq Int_g(Cl(A))$.Conversely, suppose that $A \subseteq Int_g(Cl(A))$. So $X \setminus Int_g(Cl(A)) \subseteq X \setminus A$, therefore $Int_g(Cl(X \setminus A)) \subseteq X \setminus A$, then $X \setminus A$ is Hpre-closed, so A is Hpre-open.

Lemma 1.4.5. If there exists g-open set U such that $U \subseteq A \subseteq Int_g(Cl(U))$, then A is $Hpre$-open.

Proof.

Directly from Lemma 1.4.2 and Lemma 1.3.2.

Remark 1.4.6. The opposite of above lemma is not true. In Example (1.3.4) note that, $\{x_1, x_2\} \in HpreO(X)$and $\{x_2\} \in GO(X)$, but $\{x_2\} \subset \{x_1, x_2\} \not\subset \{x_2\}$.

Theorem 1.4.7. In any topological space $(X, \tau)$the statements are verified

(1) Any *pre*-open subset of X is *Hpre*-open set.

(2) Any g-open subset of X is *Hpre*-open set.

Proof.

Directly from Theorem 1.3.5.

Remark 1.4.8. The opposite of abovetheorem is not trueas a show from Example (1.3.4) , since $\{x_2\} \in HpreO(X)$, but $\{x_2\} \notin preO(X)$ and $\{x_1, x_3\} \in HpreO(X)$, but $\{x_1, x_3\} \notin GO(X)$. The above outcomes can be outlined in the accompanying chart, however the opposite is not true.

$$pre\text{-open set} \rightarrow Hpre\text{-open set} \longleftarrow g\text{-open set}$$

Theorem 1.4.9. If $\{F_i : i \in \Delta\}$ is a collection of *Hpre*-open sets of a space (X, τ), then $\cup_{i\in\Delta} F_i$ is *Hpre*-open set.

Proof.

Since $G_i \subseteq Int_g(Cl(G_i))$ for every $i \in \Delta$ and $G_i \subseteq \cup_{i\in\Delta} G_i$for every , $Cl(G_i) \subseteq Cl(\cup_{i\in\Delta} G_i)$. So $\cup_{i\in\Delta} Cl(G_i) \subseteq Cl(\cup_{i\in\Delta} G_i)$

Hence $\bigcup_{i\in\Delta} G_i \subseteq \bigcup_{i\in\Delta} Int_g Cl(G_i)) \subseteq Int_g(\bigcup_{i\in\Delta} Cl(G_i)) \subseteq Int_g(Cl(\bigcup_{i\in\Delta} G_i)$

Therefore $\bigcup_{i\in\Delta} G_i$is $Hpre$-open set .

Remark 1.4.10. The intersection of two $Hpre$-open sets is not necessary by $Hpre$-oen set. In Example (1.3.9) , since two $Hpre$-open sets are $\{x_1, x_2\}$ and $\{x_1, x_3\}$ but $\{x_1, x_2\} \cap \{x_1, x_3\} = \{x_1\}$ is not $Hpre$-open sets.

Corollary 1.4.11. Let (X, τ) be a topological space. Then

(1) The union of an $Hpre$-open set and a pre-open set is $Hpre$-open set.

(2) The union of an $Hpre$-open set and a g-open set is $Hpre$-open set.

Proof.

(1) Directly from Theorem 1.4.7 (1) and Theorem 1.4.9.

(2) Directly from Theorem 1.4.7 (2) and Theorem 1.4.9.

Definition 1.4.12. A subset A of a space X. The $Hpre$-interior of A is denoted by $Int^H_{pre}(A)$ is the union of all $Hpre$-open sets in X contained in A. That is $Int^H_{pre}(A) = \cup \{F: A \subseteq F, F \in HpreO(X)\}$.

Lemma1.4.13. A subset A of a space X, then

(1)$X \setminus Cl^H_{pre}(A) = Int^H_{pre}(X \setminus A)$.

(2)$X \setminus Int^H_{pre}(A) = Cl^H_{pre}(X \setminus A)$.

Proof.

Easy.

Theorem 1.4.14. Let A be a subset of X. Then A is $Hpre$-open set if and only if $Int_{pre}^{H}(A) = A$.

Proof.

Directly from Theorem 1.3.13 and Lemma 1.4.13.

Theorem 1.4.15. Let A and B be a subsets of . Then the following results hold .

(1)$Int_{pre}(A) \subseteq Int_{pre}^{H}(A) \subseteq A$ and $Int_{g}(A) \subseteq Int_{pre}^{H}(A)$.

(2)$Int_{pre}^{H}(\emptyset) = \emptyset$and$Int_{pre}^{H}(X) = X$.

(3) $Int_{pre}^{H}\left(Int_{pre}^{H}(A)\right) = Int_{pre}^{H}(A)$.

(4) If $\subseteq B$, then $Int_{pre}^{H}(A) \subseteq Int_{pre}^{H}(B)$.

(5) $Int_{pre}^{H}(A) \cup Int_{pre}^{H}(B) \subseteq Int_{pre}^{H}(A \cup B)$.

(6) $Int_{pre}^{H}(A \cap B) \subseteq Int_{pre}^{H}(A) \cap Int_{pre}^{H}(B)$.

Proof.

(1) Directly from Theorem 1.4.7 (1) and (2) , respectively . Note that in general $\nsubseteq Int_{pre}^{H}(A)$. For Example 1.3.4 consider let= $\{x_3\}$, $Int_{pre}^{H}(A) = \emptyset$ so $A \subseteq Int_{pre}^{H}(A)butInt_{pre}^{H}(A) \nsubseteq (A)$.

(2) Since X and $\emptyset$ are $Hpre$- open sets then by Theorem 1.4.14 is satisfied.

(3) Since $Int_{pre}^{H}(A)$is $Hpre$ open-sets then by Theorem 1.4.14, we have $Int_{pre}^{H}\left(Int_{pre}^{H}(A)\right) = Int_{pre}^{H}(A)$.

(4) Since $Int^H_{pre}(A) \subseteq A \subseteq B$, then $Int^H_{pre}(A) \subseteq B$,therefore by (3) $Int^H_{pre}(Int^H_{pre}(A)) \subseteq Int^H_{pre}(B)$. So $Int^H_{pre}(A) = Int^H_{pre}(B)$. Note that the opposite direction of (4) is not true . For Example in 1.3.4 $A \nsubseteq B, then\ Int^H_{pre}(A) \subseteq Int^H_{pre}(B)$ if $A = \{x_3\}$ and $B = \{x_2\}$ and $Int^H_{pre}(A) = \emptyset$, $Int^H_{pre}(B) = \{x_2\}$.

(5)and **(6)** Directly from (4) .

Remark 1.4.16

The equality in the statement (5) need not be true as seen from Example 1.3.9 let $A = \{x_1\}$, $B = \{x_2, x_3\}$ and $A \cup B = X$ since $Int^H_{pre}(A) = \emptyset$, $Int^H_{pre}(B) = \{x_2, x_3\}$ and $Int^H_{pre}(A \cup B) = X$, therefore$Int^H_{pre}(A \cup B) \nsubseteq Int^H_{pre}(A) \cup Int^H_{pre}(B)$.

Remark 1.4.17

. The equality in the statement (6) need not be true as seen from example in 1.3.9 let $A = \{x_1, x_2\}$, $B = \{x_1, x_3\}\ and\ A \cap B = \{x_1\}$ since $Int^H_{pre}(A \cap B) = \emptyset$, $Int^H_{pre}(A) = \{x_1, x_2\}$ and $Int^H_{pre}(B) = \{x_1, x_3\}$, therefore $Int^H_{pre}(A) \cap Int^H_{pre}(B) \nsubseteq Int^H_{pre}(A \cap B)$.

Proposition 1.4.18. Any a subset $A$of (X, τ), then

(1)$A \cap Int_g(Cl(A))$is $Hpre$-open set.

(2) $A \cup Cl_g(Int(A))$is $Hpre$-closed set.

Proof.

(1)$Int_g(Cl(A \cap Int_g(Cl(A)))) = Int_g\left(Cl(A) \cap Int_g(Cl(A))\right) = Int_g(Cl(A))$. Then $A \cap Int_g(Cl(A)) = A \cap Int_g(Cl(A \cap Int_g(Cl(A)))) \subseteq Int_g(Cl\left(A \cap Int_g(Cl(A))\right))$. So $A \cap Int_g(Cl(A))$ is $Hpre$-open set.

(2) From (1) we have $X \setminus (A \cup Cl_g(Int(A)) = (X \setminus A) \cap Int_g(Cl(X \setminus A))$ is $Hpre$-open set. This also means $A \cup Cl_g(Int(A))$ is $Hpre$-closed set.

Proposition 1.4.19. A subset A of a space (X, τ), then

(1) $Int^H_{pre}(A) = A \cap Int_g(Cl(A))$.

(2) $Cl^H_{pre}(A) = A \cup Cl_g(Int(A))$.

Proof.

(1) Suppose that $M = Int^H_{pre}(A)$. It is clear that M is $Hpre$-open set and $M \subseteq A$. Since M is $Hpre$-open set, $M \subseteq Int_g(Cl(M)) \subseteq Int_g(Cl(A)$. This proves it $M \subseteq A \subseteq Int_g(Cl(M))$. By Proposition 1.4.16, $A \cap Int_g(Cl(A))$ is $Hpre$-open set. By the definition of $Int^H_{pre}(A)$, $A \cap Int_g(Cl(A)) \subseteq M$. Therefore $M = A \cap Int_g(Cl(A))$. So $Int^H_{pre}(A) = A \cap Int_g(Cl(A))$.

(2) By Lemma 1.4.13 we will get, $Cl^H_{pre}(A) = X \setminus Int^H_{pre}(X \setminus A) = X \setminus ((X \setminus A) \cap Int_g(Cl(A)))$, by (1) $= X \setminus (X \setminus A) \cup \left(X \setminus Int_g(Cl(X \setminus A))\right) = A \cup Cl_g(Int(A))$.

Theorem 1.4.20. In any topological space (X, τ). Let $\in X$, then $x \in Cl^H_{pre}(A)$ if and only if for every $Hpre$-open set U containing x and A is not empty .

Proof.

⇒) Suppose that $x \in Cl_{pre}^{H}(A)$ and there exists $Hpre$-open set U containing x such that $\cap U = \emptyset$. Therefore $A \subseteq X\backslash U$ and $X\backslash U$ is $Hpre$-closed set .Then $Cl_{pre}^{H}(A) \subseteq Cl_{pre}^{H}(X\backslash U) = X\backslash U$. So that $x \notin Cl_{pre}^{H}(A)$, which is a contradiction .

⇐) Assume that $U \cap A \neq \emptyset$ for each $Hpre$-open set U containing $x$and $x \notin Cl_{pre}^{H}(A)$. Therefore there exists $Hpre$-closed subset G containing A such that $\notin G$. Then $x \in X\backslash G$ and $X\backslash G$ is $Hpre$-open . So that $A \subseteq G$ and $(X\backslash G\ ) \cap A$ is empty . This is a contradiction to our claim.

1.5.$Hpre$ – Neighbourhood

Now let us introduce the $Hpre$ – neighbourhood and we study some of its properties .

Definition 1.5.1. Let (X, τ) be a topological space .A subset G of X is $Hpre$- neighbourhood of point $x \in X$, if and only if there exists an $Hpre$–open set A such that $x \in A \subseteq G$. Also G is called $Hpre$ – neighbourhood of $\subseteq X$, if and only if there exists an $Hpre$ –open set A , such that $B \subset A \subseteq G$.

Remark 1.5.2. Form the above definition we get that the any $Hpre$ –open set is a $Hpre$-neighbourhood . But the converse is not true . In Example 1.3.4 , note that $\{x_2, x_3\}$ is a $Hpre$ – neighbourhood of the set $\{x_2\}$, but it is not $Hpre$-open set .

Theorem 1.5.3 . Let (X, τ) be a topological space and for any $\in X$, let $N_{Hpre}(x)$ be the collection of all $Hpre$ –neighbourhood of $x \in X$. Now we give some properties of system of $Hpre$-neighbourhood.

(1) For any $x \in X, N_{Hpre}(x) \neq \emptyset$.

(2) Every $Hpre$ - neighbourhood G of x , contains x.

(3) If $G \in N_{Hpre}(x)$ and $G \subseteq F$, then $F \in N_{Hpre}(x)$.

(4) Let $x \in X$ and F is a neighbourhood of x and G is $Hpre$-neighbourhood of x . Then $F \cap G$ is $Hpre$ - neighbourhood of the point x.

(5) If $\in N_{Hpre}(x)$, then there exists $F \in N_{Hpre}(x)$, such that $F \subset G$ and $F \in N_{Hpre}(y)$, for any $y \in F$.

Proof .

(1) Since X is $Hpre$-open set. Therefore X is $Hpre$-neighbourhood of any point $\in X$. So $N_{Hpre}(X)$ is not empty.

(2) By the definition of $Hpre$-neighbourhood we can get that every point $x \in X$ belongs to its $Hpre$-neighbourhood .

(3) Let $\in N_{Hpre}(x)$, then there exists a $Hpre$-open set A , such that $x \in A \subseteq G$. Therefore since $G \subset F$ then $x \in A \subseteq G \subset F$ is also $Hpre$-neighbourhood of the point x.

(4) Let F be a neighbourhood and G be a $Hpre$-neighbourhood of $x \in X$. therefore there exist open setA_1 and $Hpre$-open set A_2such that $x \in A1 \subseteq F$and $\in A_2 \subseteq G$. Therefore $x \in A_1 \cap A_2 \subseteq F \cap G$. And since $A_1 \cap A_2$ is $Hpre$-

open set , therefore $F \cap G$ is $Hpre$-neighbourhood of the point x .

(5) Let $G \in N_{Hpre}(x)$ so there exists a $Hpre$-open set A such that $\in A \subseteq G$. Since A is a $Hpre$-open set so it is a $Hpre$-neighbourhood of any of its points , therefore $A \in N_{Hpre}(x)$ and $A \in N_{Hpre}(y)$, for each $y \in A$.

Remark 1.5.4. In any topological space (X, τ) . If G and F are $Hpre$-neighbourhoods of point $x \in X$, then $G \cap F$ need not be $Hpre$-neighbourhood of x . See Remark 1.4.10.

1.6.*Hpre*- Continuous Functions

In this section we present $Hpre$-continuous functions, $Hpre$-open functions and $Hpre$-closed functions with some relations, propositions and theorems about them.

Definition 1.6.1.A function $f: X \to Y$ is said to be $Hpre$-continuous if $f^{-1}(G)$ is $Hpre$-open for every open set Gin Y.

Theorem 1.6.2. For A function $f: X \to Y$, then the following properties are verified:

(1) Any pre-continuous function is $Hpre$-continuous function.

(2) Any g-continuous function is $Hpre$-continuous function.

Proof.

Directly from Theorem 1.4.7.

Remark 1.6.3. The converse of above theorem is not trueas we see in the following example.

Example 1.6.4.(1)Let $X = \{x_1, x_2, x_3\}$ with the topology $\tau = \{X, \phi, \{x_1\}\}$ and $Y = \{y_1, y_2, y_3\}$ with the topology $\sigma = \{Y, \phi, \{y_1, y_2\}, \{y_3\}\}$. And $f: X \to Y$ be a function defined by $f(x_1) = f(x_2) = y_1$, $f(x_3) = y_3$. Now $\tau^c = \{X, \phi, \{x_2, x_3\}\}$,$preC(X) = \{X, \phi, \{x_2\}, \{x_3\}, \{x_2, x_3\}\}$, $preO(X) = \{X, \phi, \{x_1\}, \{x_1, x_2\}, \{x_1, x_3\}\}$, $GC(X) = \{X, \phi, \{x_2\}, \{x_1, x_3\}, \{x_3\}, \{x_1, x_2\}, \{x_2, x_3\}\}$, $GO(X) = \{X, \emptyset, \{x_1\}, \{x_3\}, \{x_2\}, \{x_1, x_3\}, \{x_1, x_2\}\}$, $HpreC(X) = HpreO(X) = P(X)$.Since $\{y_3\}$ is open in Y, $f^{-1}(\{y_3\}) = \{x_3\} \in HpreO(X)$, but $\{x_3\} \notin preO(X)$. Then f is $Hpre$-continuous function, but not pre-continuous function.

(2) Let (X, τ) and (Y, σ) be two topological spaces in (1) and $f: X \to Y$ be a function defined by $f(x_1) = y_1$, $f(x_2) = f(x_3) = y_3$. Note that $\{y_3\}$ is open in Y, $f^{-1}(\{y_3\}) = \{x_2, x_3\} \in HpreO(X)$,

but$\{x_2, x_3\} \notin GO(X)$. Then f is $Hpre$-continuous function, but not g-continuous function.

The above outcomes can be abridged in the accompanying chart, yet the oppositeis not true.

pre-continuous function→ $Hpre$-continuous function← g-continuous function

Theorem 1.6.5. A function $f: X \to Y$ is $Hpre$-continuous function iff satisfying one of the following properties:

(1) For every $x \in X$ and any open set $G \subset Y$ containing $f(x)$, there exists $Hpre$-open set $A \subset X$ containing x such that $f(A) \subset G$.

(2) The inverse image of every closed subset F in Y is $Hpre$-closed in X.

(3) $Cl_{pre}^{H}(f^{-1}(B)) \subseteq f^{-1}(Cl(B)), B \subseteq Y$.

(4) $f^{-1}(Int(B)) \subseteq Int_{pre}^{H}(f^{-1}(B)), B \subseteq Y$.

Proof.

(1) $\Rightarrow$) Suppose that f is $Hpre$-continuous . Let x be any point of X and G any open set containing $f(x)$ in Y .Then $f^{-1}(G) \in HpreO(X)$. The set $A = f^{-1}(G)$ which contains x , so that $(A) \subset G$.

$\Leftarrow$) Let G be any opensubset of Y. Suppose that $\in f^{-1}(G)$, then $f(x) \in G$. And let A be any $Hpre$-open subset of , such that $x \in A$ and $f(A) \subset G$, but $A \subset f^{-1}(G)$. Hence $\in A \subset f^{-1}(G)$, for each $x \in f^{-1}(G)$, therefore $f^{-1}(G)$ is the union of open subset of X and thus , an $Hpre$-open subset . Then f is $Hpre$-continuous .

(2) $\Rightarrow$)Suppose that f is $Hpre$-continuous. To proof $X \setminus f^{-1}(F) \in HpreO(X)$, for every $Y \setminus F \in Y$. Let F is closed set in Y. This implies $Y \setminus F$ is open in $Y$and $f^{-1}(Y \setminus F) \in HpreO(X)$.

But $f^{-1}(Y \setminus F) = X \setminus f^{-1}(F)$. Then $X \setminus f^{-1}(F) \in HpreO(X)$. So $f^{-1}(F)$ is $Hpre$-closed in X.

$\Leftarrow$) Suppose that $X \setminus f^{-1}(F) \in HpreO(X)$, for every $Y \setminus F \in Y$. To proof f is $Hpre$-continuous, let V is open in Y so $Y \setminus V$ is closed in Y this imples $f^{-1}(Y \setminus F) \in HpreC(X)$.Then $X \setminus f^{-1}(Y \setminus V) \in HpreO(X)$. But, $X \setminus f^{-1}(Y \setminus V) = X \setminus \left(X \setminus f^{-1}(V)\right) = f^{-1}(V) \in HpreO(X)$. Therefore f is $Hpre$-continuous.

(3)$\Rightarrow$ Suppose that f is $Hpre$-continuous. Let $B \subseteq Y$, $B \subseteq Cl(B)$ then $f^{-1}(B) \subseteq f^{-1}(Cl^{H}_{pre}(B))$. Since $Cl(B)$ is a closed set in Y by (2) $f^{-1}(Cl^{H}_{pre}(B)) \in HpreC(X)$ and therefore $Cl^{H}_{pre}(f^{-1}(B)) = f^{-1}(Cl(B))$. Now $B \subseteq Cl(B)$ implies that $f^{-1}(B) \subseteq f^{-1}(Cl(B))$so $Cl^{H}_{pre}(f^{-1}(B)) \subseteq Cl^{H}_{pre}(f^{-1}(Cl(B))) = f^{-1}(Cl(B))$.

$\Leftarrow$) Let the condition hold and let F be any closed set in Y so that $Cl(F) = F$. By hypothesis $Cl^{H}_{pre}(f^{-1}(F)) \subseteq f^{-1}\left(Cl(F)\right) = f^{-1}(F)$. But, $f^{-1}(F) \subseteq Cl^{H}_{pre}(f^{-1}(F))$and $Cl^{H}_{pre}(f^{-1}(F)) = f^{-1}(F)$. Hence $f^{-1}(F) \in HpreC(X)$. It follows from (2)that$f$ is $Hpre$-continuous.

(4)$\Rightarrow$) Let f is $Hpre$-continuous. Since $Int(B)$ is open in Y, $f^{-1}(Int(B)) \in HpreO(X)$ and therefore $Int^{H}_{pre}(f^{-1}(Int(B)) = f^{-1}(Int(B))$. Now $Int(B) \subseteq B$ this implies $f^{-1}(Int(B)) \subseteq f^{-1}(B)$therefore $Int^{H}_{pre}(f^{-1}(Int(B))) \subseteq Int^{H}_{pre}(f^{-1}(B))$, so $f^{-1}(Int(B)) \subseteq Int^{H}_{pre}(f^{-1}(B))$.

$\Leftarrow$) Let the condition hold and let G be any open set in Y, so that $Int(G) = G$. By hypothesis $f^{-1}(Int(G)) \subseteq Int^{H}_{pre}(f^{-1}(G))$. But,

$Int^H_{pre}(f^{-1}(G)) \subseteq f^{-1}(G)$, so $Int^H_{pre}(f^{-1}(G)) = f^{-1}(G)$. Therefore $f^{-1}(G) \in HpreO(X)$. Then f is $Hpre$-continuous.

Let us now use $Hpre$-open sets to define of $Hpre$-open function . Also we introduce some theorems about these notions .

Definition 1.6.6.A function $f: X \to Y$ is called$Hpre$-open (resp. $Hpre$-closed) if $f(G)$ is $Hpre$-open (resp. $Hpre$-closed) in Y for every open (resp. closed) set Gin X.

Theorem 1.6.7. For A function $f: X \to Y$, then the following properties are verified:

(1) Any pre-open function is $Hpre$-open function.
(2) Any g-open function is $Hpre$-open function.
Proof.

Directly from Theorem 1.4.7.

Remark 1.6.8.(1) $Hpre$-open function does not have to be pre-open function (see Example 6.9(1)).
(2)$Hpre$-open function does not have to be g-open function (see Example 6.9(2)).

Example 1.6.9.(1)Let $X = \{x_1, x_2, x_3\}$ with the topology $\tau = \{X, \phi, \{x_2\}\}$ and $Y = \{y_1, y_2, y_3\}$ with the topology $\sigma = \{Y, \phi, \{y_2, y_3\}, \{y_2\}\}$. A function $f: X \to Y$ defined by $f(x_1) = y_2$,

$f(x_2) = y_3$ and $f(x_3) = y_1$. Now $\sigma^c = \{Y, \phi, \{y_1\}, \{y_1, y_3\}\}$,$preO(Y) = \{Y, \phi, \{y_2\}, \{y_1, y_2\}, \{y_2, y_3\}\}$, $preC(Y) = \{Y, \phi, \{y_1\}, \{y_3\}, \{y_1, y_3\}\}$, $GO(Y) = \{Y, \phi, \{y_2\}, \{y_3\}, \{y_2, y_3\}\}$, $GC(Y) = \{Y, \phi, \{y_1\}, \{y_1, y_2\}, \{y_1, y_3\}\}$, $HpreO(Y) = \{Y, \phi, \{y_2\}, \{y_3\}, \{y_1, y_2\}, \{y_2, y_3\}\}andHpreC(Y) = \{Y, \phi, \{y_1\}, \{y_3\}, \{y_1, y_2\}, \{y_1, y_3\}\}$ Since $\{x_2\}$ is open in X, $f(\{x_2\}) = \{y_3\} \in HpreO(Y)$, but $\{y_3\} \notin preO(Y)$. Then f is *Hpre*-open function, but not *pre*-open function.

(2) Let $X = \{x_1, x_2, x_3\}$ with the topology $\tau = \{X, \phi, \{x_1, x_3\}\}$ and $Y = \{y_1, y_2, y_3\}$ with the topology $\sigma = \{Y, \phi, \{y_1, y_2\}\}$.

A function $f: X \to Y$ be a function defined by $f(x_1) = y_2$, $f(x_2) = y_1$ and $f(x_3) = y_3$. Now $\sigma^c = \{Y, \phi, \{y_3\}\}$, $GO(Y) = \{Y, \phi, \{y_1\}, \{y_2\}, \{y_1, y_2\}\}$, $GC(Y) = \{Y, \phi, \{y_3\}, \{y_1, y_3\}, \{y_2, y_3\}\}$, $preO(Y) = HpreO(Y) = \{Y, \phi, \{y_1\}, \{y_2\}, \{y_1, y_3\}, \{y_1, y_2\}, \{y_2, y_3\}\}$ and $preC(Y) = HpreC(Y) = \{Y, \phi, \{y_1\}, \{y_3\}, \{y_2\}, \{y_1, y_3\}, \{y_2, y_3\}\}$ Since $\{x_1, x_3\}$ is open in X, $f(\{x_1, x_3\}) = \{y_2, y_3\} \in HpreO(Y)$, but $\{y_2, y_3\} \notin preO(Y)$. Then f is *Hpre*-open function, but not g-open function.

The above outcomes can be condensed in the accompanying chart, yet the oppositeis not true.

pre-open function→ *Hpre*-open function⟵ g-open function

Theorem 1.6.10. For a function $f: X \to Y$, then the following properties are equivalent:

(1) f is $Hpre$-open function.

(2) For every $x \in X$ and every neighbourhood V of x, there exists $Hpre$-open set $U \subseteq Y$ containing $f(x)$ such that $U \subseteq f(V)$.

Proof.

(1)⇒(2) Let $x \in X$ and V is a neighbourhood of x, therefore there exists open set $W \subseteq X$ such that $x \in W \subseteq V$. A set $U = f(W)$. Since f is $Hpre$-open function, $f(W) = U \in HpreO(Y)$ and so $f(x) \in U \subseteq f(V)$.

(2)⇒(1) Clearly.

Theorem 1.6.11. Let $f: X \to Y$ be $Hpre$-closed (resp. $Hpre$-open) function and $U \subseteq Y$. If $A \subseteq X$ is an open (resp. closed) set containing $f^{-1}(U)$, then there exists $Hpre$-open (resp. $Hpre$-closed) set $V \subseteq Y$ containing U such that $f^{-1}(V) \subseteq A$.

Proof.

Suppose that $U \subseteq Y$ and let A be an open (resp. closed) subset of X such that $f^{-1}(U) \subseteq A$. Then $V = Y \setminus f(X \setminus V)$ is $Hpre$-open (resp. $Hpre$-closed) set containing U and $f^{-1}(V) \subseteq A$.

Corollary 1.6.12. For every $A \subseteq Y$ and $f: X \to Y$ is $Hpre$-open function, then $f^{-1}(Cl^{H}_{pre}(A)) \subseteq Cl(f^{-1}(A))$.

Proof.

Since $Cl(f^{-1}(A))$ is closed in X containing $f^{-1}(A)$ for any set $A \subseteq Y$. By Theorem 1.6.11, then there exists $Hpre$-closed set $V \subseteq Y,\ A \subseteq V$ such that $f^{-1}(V) \subseteq Cl(f^{-1}(A))$. Therefore $f^{-1}(Cl^H_{pre}(A)) \subseteq f^{-1}(Cl^H_{pre}(V)) \subseteq f^{-1}(V) \subseteq Cl(f^{-1}(A))$.

Theorem 1.6.13. A function $f: X \to Y$ is $Hpre$-open if and only if $f(Int(B)) \subseteq Int^H_{pre}(f(B))$ for any subset $B$of X.

Proof.

Suppose that $f: X \to Y$ is $Hpre$-open function and $B \subseteq X$. Then $Int(B)$ is open set in X and $f(Int(B))$ is $Hpre$-open set contained in $f(B)$. Therefore $f(Int(B)) \subseteq Int^H_{pre}(f(B))$. Conversely, let $f(Int(B)) \subseteq Int^H_{pre}(f(B))$ for any subset B of X and V is open set in X. Then $Int(V) = V,\ f(V) \subseteq Int^H_{pre}(f(V))$. Therefore $f(V) = Int^H_{pre}(f(V))$. By Theorem 1.4.14 $f(V)$ is $Hpre$-open function.

Theorem 1.6.14. A function $f: X \to Y$ is $Hpre$-closed if and only if $Cl^H_{pre}(f(A)) \subseteq f(Cl(A))$.

Proof.

Let f is a $Hpre$-closed function and $A \subseteq X$. Since $Cl(A)$ is closed set in X. Then $f(Cl(A))$ is $Hpre$-closed set and consequently $f\big(Cl(A)\big) = Cl^H_{pre}(f(Cl(A)))$. Since $A \subseteq Cl(A)$ this implies $Cl^H_{pre}(f(A)) \subseteq Cl^H_{pre}(f(Cl(A)))$, therefore $Cl^H_{pre}(f(A)) \subseteq f(Cl(A))$. Conversely, let $Cl^H_{pre}(f(A)) \subseteq f(Cl(A))$, for every

$A \subseteq X$ and let H is a closed set so that $Cl(H) = H$. Hence $Cl_{pre}^{H}(f(H)) \subseteq f(Cl(H)) = f(H)$but $f(H) \subseteq Cl_{pre}^{H}(f(H))$. Therefore $f(H) = Cl_{pre}^{H}(f(H))$. Hence f is $Hpre$-closed function.

Theorem 1.6.15. If $f: X \rightarrow Y$ is bijective then the following statements are equivalent:

(1)$f^{-1}$is$Hpre$-continuous function.

(2) $f$is$Hpre$-open function.

(3) $f$is $Hpre$-closed function.

Proof.

(1)⇒(2) Let G is open in X. Then $X \setminus G$ is closed in X. Since f^{-1} is $Hpre$-cntinuous function, $(f^{-1})^{-1}(X \setminus G)$ is $Hpre$-closed in Y. This implies $f(X \setminus G) = Y \setminus f(G)$ is $Hpre$-closed in Y. Then $f(G)$ is $Hpre$-open in Y. Therefore f is $Hpre$-open function.

(2)⇒(3) Assume that that B is closed set in X. So $(X \setminus B)$ is open in X. Since f is $Hpre$-open function, $f(X \setminus B)$ is $Hpre$-open in Y. Therefore $f(X \setminus B) = Y \setminus f(B)$ is $Hpre$-open in Y. So $f(B)$ is $Hpre$-closed in Y. Then f is $Hpre$-closed function.

(3)⇒(1) Assume that B is closed set in X. Since f is $Hpre$-open function, $f(B)$ is $Hpre$-closed. This mean $(f^{-1})^{-1}(B)$ is $Hpre$-closed in Y. Then $f^{-1}$is $Hpre$-continuous function.

Remark 1.6.16. If $f: X \rightarrow Y$ and If$g: Y \rightarrow Z$ are both $Hpre$-open functions, then the composition $gof(x)$ need not is$Hpre$-open function seen the following example.

Example 1.6.17. Let $X = \{x_1, x_2, x_3\}$ with the topology $\tau = \{X, \phi, \{x_3\}\}$, $Y = \{y_1, y_2, y_3\}$ with the topology $\sigma = \{Y, \phi, \{y_1\}, \{y_1, y_2\}\}$ and $Z = \{z_1, z_2, z_3\}$with the topology $\rho = \{Z, \emptyset, \{z_1, z_2\}\}$. A function $f: X \to Y$ defined by

$f(x_1) = y_1$, $f(x_2) = y_3$ and $f(x_3) = y_2$ and $g: Y \to Z$ is defined by $g(y_1) = z_1$, $g(y_2) = z_3$ and $f(y_3) = z_2$. Since $\sigma^c = \{Y, \phi, \{y_3\}, \{y_2, y_3\}\}$,$GO(Y) = \{Y, \phi, \{y_1\}, \{y_2\}, \{y_1, y_2\}\}$, $GC(Y) = \{Y, \phi, \{y_3\}, \{y_1, y_3\}, \{y_2, y_3\}\}$, $HpreO(Y) = \{Y, \phi, \{y_1\}, \{y_1, y_2\}, \{y_2\}, \{y_1, y_3\}\}$, $HpreC(Y) = \{Y, \phi, \{y_2\}, \{y_3\}, \{y_1, y_3\}, \{y_2, y_3\}\}$ and $\rho^c = \{Z, \phi, \{z_3\}\}$,$GO(Z) = \{Z, \phi, \{z_1\}, \{z_2\}, \{z_1, z_2\}\}$, $GC(Z) = \{Z, \phi, \{z_3\}, \{z_1, z_3\}, \{z_2, z_3\}\}$, $HpreO(Z) = \{Z, \phi, \{z_1\}, \{z_2\}, \{z_1, z_2\}, \{z_1, z_3\}, \{z_2, z_3\}\}$, $HpreC(Z) = \{Z, \phi, \{z_1\}, \{z_3\}, \{z_2\}, \{z_2, z_3\}, \{z_1, z_3\}\}$. It is clear that, f and g are *Hpre*-open function. Since $\{x_3\}$ is open in X. But $gof(\{x_3\}) = g(\{y_2\}) = \{z_3\}$ which is not *Hpre*-open in Z. Then gof is not *Hpre*-open function.

Chapter Two

Applications of *Hpre*-Closed Sets in The Functions

2.1. Introduction

In this chapter we introduce the associative separation axioms of the *Hpre*-open sets and introduce the *Hpre*-closed graph and strongly *Hpre*-closed by *Hpre*-closed sets which are defined in the previous chapter and then give some new theorems about them .

2.2. T_i-Spaces with Respect to *Hpre*-Open Sets , for i=1,2

Let us now introduce our definitions related to the T_1 and T_2 spaces , with some theorems and relations .

Definition 2.2.1. Let (X,τ) be a topological space. Then the space (X,τ) is called $Hpre-T_1$ space iff for every distinct pair of points x_1 and x_2 in X, there exists *Hpre*-open sets U in X containing x_1 but not x_2, and a *Hpre*-open set V in X containing x_2 but not x_1.

Theorem 2.2.2. In any topological space $(X,\tau)$the statements are verified

(1) Any *pre*-T_1 space is $Hpre-T_1$ space.

(2) Any g-T_1 space is $Hpre-T_1$ space.

Proof.

It is clear that from Theorem 1.4.7.

Remark 2.2.3. **(1)**$Hpre-T_1$space does not have to be *pre*-T_1 space(see Example 1.6.4(1)).

(2) $Hpre-T_1$space does not have to be g-T_1 space (see Example 1.3.9).

The above results can be summarized in the following diagram , but the convers is not true

$$\text{pre}-\text{T}_1\text{space} \rightarrow \text{Hpre}-\text{T}_1\text{space} \leftarrow \text{g}-\text{T}_1 \text{ space}$$

Theorem 2.2.4. Let (X,τ) be a topological space. Then

(1) $X$is $Hpre-T_1$ space if and only if for each $x_1 \neq x_2 \in X$, $\{x_1\}$ is *pre*-closed set and $\{x_2\}$ is *Hpre*-closed set in X.

(2) $X$is $Hpre-T_1$ space if and only if for each $x_1 \neq x_2 \in X$, $\{x_1\}$ is g-closed set and $\{x_2\}$ is *Hpre*-closed set in X.

Proof.

(1) Suppose that X is a $Hpre-T_1$ space and let $x_1 \neq x_2 \in X$ then there exists a *Hpre*-open set U_{x_1} containingx_1 but not x_2, and a *Hpre*-open set U_{x_2} containing x_2 but not x_1. Then $x_1 \in U_{x_1} \subset \{x_2\}^c$. Therefore $\{x_2\}^c = \cup_{x_{1\in\{x_2\}^c}} U_{x_1}$, which is *Hpre*-open set and $\{x_2\}$ is *Hpre*-closed set in X. Also$\{x_1\}$ *Hpre*-closed set in X. In fact $x_2 \in U_{x_2} \subset \{x_1\}^c$, then $\{x_1\}^c = \bigcup_{x_2 \in \{x_1\}^c} U_{x_2}$, which is $Hpre$-open set and $\{x_1\}$ is $Hpre$-closed set in X. Now for the converse, let $x_1 \neq x_2 \in X$, $U_{x_1} = X \setminus \{x_2\}$ is$Hpre$-open set in X containing x_1 but not x_2, and $U_{x_2} = X \setminus \{x_1\}$ is is $Hpre$-open set in X containing x_2 but not x_1, therefore X is $Hpre - T_1$ space.

(2) Similar proof of (1).

Definition 2.2.5. Let (X, τ) be a topological space. Then the space (X, τ) is called $Hpre - T_2$ space iff for every distinct pair of points x_1 and x_2 in X, there exists $Hpre$-open sets U and V in X such that $x_1 \in U$ and $x_2 \in V$, and $U \cap V = \emptyset$.

Theorem 2.2.6. In any topological space $(X, \tau)$the statements are verified:

(1) Every *pre*-T_2 space is $Hpre - T_2$ space.

(2) Every g-T_2 space is $Hpre - T_2$ space.

Proof.

Direct that from Theorem 1.4.7.

Remark 2.2.7. (1)$Hpre - T_2$space does not have to be *pre*-T_2 space (see Example 1.6.4(1)).

(2)$Hpre - T_2$space does not have to be g-T_2 space (see Example 1.3.9).

(3) Every $Hpre - T_2$ space is $Hpre - T_1$ space but the converse is not true (see Example 1. 6.9(2)).

The above results can be summarized in the following diagram , but the convers is not true

$$\text{pre}-\text{T}_2\text{space} \rightarrow \text{Hpre} - \text{T}_2\text{space} \leftarrow \text{g} - \text{T}_2 \text{ space}$$

Theorem 2.2.8. Let (X, τ) be a topological space and let (Y, σ) be T_2-space. Let $f: X \rightarrow Y$ be a one-one and $Hpre$-continuous function. Then (X, τ) is $Hpre - T_2$ space.

Proof.

Let $x_1 \neq x_2 \in X$. Since f is one-one, $x_1 \neq x_2$then $f(x_1) \neq f(x_2)$. Let $y_1 = f(x_1)$, $y_2 = f(x_2)$, so that $x_1 = f^{-1}(y_1)$ and $x_2 = f^{-1}(y_2)$. Then $y_1 \neq y_2 \in Y$. Since Y is T_2-space, then there exists a two open set U and V such that $y_1 \in U$ and $y_2 \in V$and $U \cap V = \emptyset$. Since f is $Hpre$-continuous function, so $f^{-1}(U)$ and $f^{-1}(V)$ are $Hpre$-open set in X and $f^{-1}(U) \cap f^{-1}(V) = f(U \cap V)^{-1} = \emptyset$, $y_1 \in U$, $f^{-1}(y_1) \in f^{-1}(U)$ this implies $x_1 \in f^{-1}(U)$ and $y_2 \in V$, $f^{-1}(y_2) \in f^{-1}(V)$ this implies $x_2 \in f^{-1}(V)$. So (X, τ) is $Hpre - T_2$ space.

Theorem 2.2.9. Let X be T_2-space (resp.*pre*-T_2 space) and $f: X \to Y$ *Hpre*-open function, one-one and onto, then Y is $Hpre - T_2$ space.

Proof.

Let $y_1 \neq y_2 \in X$. Since f is one-one and onto function, there exists $x_1 \neq x_2 \in X$ such that $y_1 = f(x_1), y_2 = f(x_2)$, since X is T_2-space (resp. *pre*-T_2 space) then there exists are two open (resp. *pre*-open) sets U and V in X such that $x_1 \in U,\ x_2 \in V$ and $U \cap V = \emptyset$, since f is *Hpre*-open function, therefore $f(U)$ and $f(V)$ are *Hpre*-open sets such that $y_1 = f(x_1) \in f(U),\ y_2 = f(x_2) \in f(V)$ and $f(U) \cap f(V) = f(U \cap V) = \emptyset$. Then Y is $Hpre - T_2$ space.

2.3. *Hpre*-Closed Graph

Let us introduce *Hpre*-closed graph and strongly *Hpre*-closed and the proof some theorems and result.

Definition 2.3.1.A function $f : X \to Y$ is called *Hpre*-closed graph if for any $(x, y) \in (X \times Y) \backslash G(f)$, there exist $A \in HpreO(X, x)$ and $B \in G\,O(Y, y)$ such that $(A \times Cl_g(B)) \cap G(f) = \emptyset$.

Remark 2.3.2 Obviously every closed graph is *Hpre*-closed , but the converse is not true see example the following :

Example 2.3.3. Let $X = \{x_1, x_2, x_3\}$ with the topology $\tau = \{X, \emptyset, \{x_2\}, \{x_2, x_3\}\}$ *and* Y= $\{y_1, y_2, y_3\}$ *with the topology* $\sigma = \{Y, \emptyset, \{y_1\}, \{y_1, y_2\}\}$. *A* function$f : X \to Y$be defined by $f(x_1) = f(x_3) = y_1$ and $f(x_2) = y_2$. Now τ^c={X, $\emptyset$, $\{x_1\}$,$\{x_1, x_3\}\}$, $pre\ O(X) = \{X, \emptyset, \{x_2\}, \{x_2, x_3\}, \{x_1, x_2\}\}$, $Pre\ C(X) = \{X, \emptyset, \{x_1\}, \{x_3\}, \{x_1, x_3\}\}$, $GO(X) = \{X, \emptyset, \{x_2\}, \{x_2, x_3\}, \{x_3\}\}$,

$G\ C(X) = \{X, \emptyset, \{x_1\}, \{x_1, x_2\}, \{x_1, x_3\}\}$, $HpreO(X) = \{X, \emptyset, \{x_2\}, \{x_3\}, \{x_2, x_3\}, \{x_1, x_2\}\}$, $HpreC(X) = \{X, \emptyset, \{x_1\}, \{x_3\}, \{x_1, x_2\}, \{x_1, x_3\}\}$, $\sigma^c = \{Y, \emptyset, \{y_3\}, \{y_2, y_3\}\}$, $pre\ O(Y) = \{Y, \emptyset, \{y_1\}, \{y_1, y_2\}, \{y_1, y_3\}\}$, $pre\ C(Y) = \{Y, \emptyset, \{y_2\}, \{y_3\}, \{y_2, y_3\}\}$, $GO(Y) = \{Y, \emptyset, \{y_1\}, \{y_2\}, \{y_1, y_2\}\}$, $GC(Y) = \{Y, \emptyset, \{y_3\}, \{y_2, y_3\}, \{y_1, y_3\}\}$, $HpreO(Y) = \{Y, \emptyset, \{y_1\}, \{y_2\}, \{y_1, y_2\}, \{y_1, y_3\}\}$, $HpreC(Y) = \{Y, \emptyset, \{y_2\}, \{y_3\}, \{y_1, y_3\}, \{y_2, y_3\}\}$. *Since* $\{x_1, x_2\} \in HpreO(X, x_1)$ *and* $\{y_2\} \in GO(Y, y_2)$, *but* $\{\{x_1, x_2\} \notin O(X)$ *and* $\{y_2\} \notin O(Y)$. *Then* $G(f)$ is $Hpre$ - closed but not closed .

Theorem 2.3.4.A function $f : X \to Y$ is $Hpre$ –closed graph if and only if for each $(x, y) \in (X, Y) \setminus G(f)$, there exist $A \in HpreO(X, x)$ and $B \in GO(Y, y)$ such that $f(A) \cap Cl_g(B) = \emptyset$.

Proof.

Suppose that f is $Hpre$-closed graph . Then for any $(x,y) \in (X \times Y)\backslash G(f)$, there exists $A \in HpreO(X,x)$ and $B \in GO(Y,y)$ such that $(A \times Cl_g(B)) \cap G(f) = \emptyset$ .therefore for each $f(x) \in f(A)$ and $y \in Cl_g(B)$. Since $y \neq f(x), f(A) \cap Cl_g(B) = \emptyset$. Now , for the converse , let $(x,y) \in (X \times Y)\backslash G(f)$. This implies that there exists $A \in HpreO(X,x)$ and $B \in GO(Y,y)$ such that $f(A) \cap Cl_g(B) = \emptyset$. Therefore $f(x) \neq y$ for any $x \in A$ and $y \in Cl_g(B)$. Hence $(A \times Cl_g(B)) \cap G(f) = \emptyset$.

Theorem 2.3.5. A function $f : X \to Y$ is $Hpre$-closed graph if for each $(x,y) \in (X \times Y)\backslash G(f)$, there exist $A \in HpreO(x,y)$ and $B \in HpreO(Y,y)$ such that $(A \times Cl^H_{pre}(B)) \cap G(f) = \emptyset$.

Proof.

Let f is $Hpre$-closed graph. Then for any $(x,y) \in (x \times y)\backslash G(f)$, there exists $A \in HpreO(X,x)$ and $B \in GO(Y,y)$ such that $(A\, X\, Cl_g(V)) \cap G(f) = \emptyset$ .Since g-open set is $Hpre$-open, $Cl^H_{pre}(B) \subseteq Cl_g(B)$. Hence$(AX\, Cl^H_{pre}(B)) \cap G(f) = \emptyset$.

Theorem 2.3.6. A function $f: X \to Y$ is $Hpre$-closed graph if for each$(x,y) \in (X \times Y)\backslash G(f)$,there exist $A \in HpreO(X,x)$ and $B \in HpreO(Y,y)$ such that $f(A) \cap Cl^H_{pre}(B) = \emptyset$.

Proof .

Suppose f is $Hpre$-closed graph then for each $(x,y) \in (X \times Y)\backslash G(f)$, there exist $A \in HpreO(X,x)$ and $B \in GO(Y,y)$ such that $f(A) \cap Cl_g(B) = \emptyset$. Since $Cl^H_{pre}(B) \subseteq Cl_g$, $f(A) \cap Cl^H_{pre}(B) \subseteq f(A) \cap Cl_g(B)=\emptyset$.

Theorem 2.3.7.A function $f: X \to Y$ be any surjection and $G(f)$ is $Hpre$-closed . Then $Y$is $g-T_1$.

Proof.

Let c and$d$ be any distinct point of Y . Since f is surjection this mean $f(x_1) = d$ for some$x_1 \in X$ and $(x_1, c) \in (X \times Y)\backslash G(f)$.Since the$Hpre$-closedness of $G(f)$ provides $A_1 \in HpreO(X, x_1)$, $B_1 \in GO(Y, c)$ such that $(A_1) \cap Cl_g(B_1) = \emptyset$. Now, $x_1 \in A_1$this implies $f(x_1) = d \in f(A_1)$. This means that $f(A_1) \cap Cl_g(B_1) = \emptyset$, warranty that $d \notin B_1$. Also from the subjectivity of f we obtain $x_2 \in X$ such that $f(x_2) = c$. Now ,$(x_2, d) \in (X \times Y)\backslash G(f)$ and the Hpre- closedness of $G(f)$ provides $A_2 \in HpreO(X, x_2)$,$B_2 \in GO(Y, d)$ such that $f(A_2) \cap Cl_g(B_2) = \emptyset$. Now $x_2 \in A_2$this implies that $f(x_2) = c \in f(A_2)$, then c $\notin B_2$therefore, we obtain sets B_1, $B_2 \in GO(Y)$

such that $c \in B_1$but $d \notin B_1$, while $d \in B_2$but $c \notin B_2$, therefore Y is $g - T_1$.

Corollary 2.3.8. Let $f: X \to Y$ be any surjection and $G(f)$*Hpre*-closed .Then Y is *Hpre*-T_1.

Proof.

Direct from Theorem 2.3.4 and Theorem 2.3.7 .

Theorem 2.3.9. A function $f: X \to Y$ be any injective and $G(f)$ *Hpre*-closed .Then X is *Hpre*-T_1 .

Proof.

Let a and b be any distinct point of$X$. Since f is injective this mean $f(a) \neq f(b)$ when of obtains that $(a, f(b)) \in (X \times Y)\backslash G(f)$.The*Hpre* -closedness of $G(f)$ provides $A_1 \in HpreO\,(x, a), B_1 \in GO(y, f(b))$ such that $f(A_1) \cap Cl_g(B_1) = \emptyset$. So $f(b) \notin f(A_1)$ this mean $b \notin A_1$ .Also $(b, f(a)) \in (X \times Y)\backslash\, G(f)$ and *Hpre*-closedness of $G(f)$ then $A_2 \in HpreO(X, b), B_2 \in GO(Y, f(a))$with $f(A_2) \cap Cl_g(B_2) = \emptyset$, warranty that $f(a) \notin f(A_2)$ and therefore $a \notin A_2$.HenceA_1 and $A_2 \in HpreO(X)$ such that $a \in A_1$but $b \notin A_1$while $b \in A_2$but $a \notin A_2$.Then X is *Hpre*-T_1.

Corollary 2.3.10. Let $f: X \to Y$ be any bijection and $G(f)$ *Hpre*-closed . Then both X and Y are *Hpre*-T_1 .

Proof.

Direct from Corollary2.3.8 and Theorem 2.3.9 .

Theorem 2.3.11. Let $f: X \to Y$ be any surjection and $G(f)$*Hpre*-closed . Then Y is g-T_2 .

Proof.

Let a and b be any distinct point of Y . Since f is surjective this mean $f(x_1) = a,$ for some$x_1 \in X$ and $(x_1, b) \in (X \times Y) \setminus G(f)$. The *Hpre*-closedness of $G(f)$ provides $A \in HpreO(X, x_1)$and $B \in GO(Y, b)$ such that $f(A) \cap Cl_g(B) = \emptyset$. Now $x_1 \in A$ this implies $f(x_1) = a \in f(A)$. This mean that $f(A) \cap Cl_g(B) = \emptyset,$ warranty that $a \notin Cl_g(B)$. Then there exists $M \in GO(Y, a)\ such\ that\ M \cap B = \emptyset\ .So\ Y\ is\ g - T_2$.

Corollary 2.3.12. A function $f: X \to Y$be any surjection and $G(f)$ Hpre-closed. Then Y is *Hpre*-T_2 .

Proof.

Direct from Theorem 2.2.6 (2) and Theorem 2.3.11 .

2.4. Strongly Hpre-Closed Graph

In this section we introduce strongly $Hpre$-closed graph and investigates some of their basic properties .

Definition 2.4.1. A function $f: X \to Y$ has a strongly $Hpre$-closed graph if for any $(x, y) \in (X \times Y) \backslash G(f)$, there exist $A \in HpreO(X, x)$ and $B \in O(Y, y)$ such that $(A \times CL(B)) \cap G(f) = \emptyset$.

Remarks 2.4.2. (1) Any strongly $Hpre$-closed graph is $Hpre$-closed , but the converse is not true (see Example2.3.3) , where $\{y_2\} \in GO(Y, y)$, but $\{y_2\} \notin O(Y)$, then $G(f)$ is $Hpre$-closed but not strongly $Hpre$-closed .

(2) Any strongly pre-closed graph (resp. strongly closed graph) is strongly $Hpre$ –closed graph , but the converse is not true.

We can summarize the relationships among the strongly closed graph by the following diagram

strongly closed ghaph→ strongly pre-closed graph →strongly $Hpre$-closed graph.

Theorem 2.4.3. A function $f: X \to Y$ is a strongly $Hpre$-closed graph if and only if for any $(x, y) \in (X \times Y) \backslash G(f)$, there exist

$A \in HpreO(X, x)$ and $B \in O(Y, y)$ such that $f(A) \cap CL(B) = \emptyset$.

Proof :

A same to the proof Theorem 2.3.4.

Theorem 2.4.4 . A function $f: X \to Y$ is a strongly $Hpre$ -closed graph if for each $(x, y) \in (X \times Y) \backslash G(f)$, there exist $A \in HpreO(X, x)$ and $B \in preO(Y, y)$ *such that* $(A \times Cl_{pre}(B)) \cap G(f) = \emptyset$.

Proof.

A same to the proof Theorem 2.3.5 .

Theorem 2.4.5. A function $f: X \to Y$ is a strongly $Hpre$ -closed graph if for each $(x, y) \in (X \times Y) \backslash G(f)$, there exist $A \in HpreO(X, x)$ and $B \in preO(Y, y)$ such that $f(A) \cap Cl_{pre}(B) = \emptyset$.

Proof.

A same to the proof Theorem 2.3.6.

Theorem 2.4.6 . Let $f: X \to Y$ is a function with a strongly $Hpre$-closed graph , then for each $x_0 \in X$, $f(x) = \cap \{Cl_{pre}(f(A)) : A \in HpreO(X, x_0)\}$.

Proof .

Suppose the Theorem is false . Then there exists $y_0 \neq f(x_0)$ such $that\ y_0 \in \cap \{Cl_{pre}(f(A) : A \in HpreO(X, x_0)$. Then $y_0 \in Cl_{pre}(f(A))$ for any $A \in HpreO(X, x_0)$. Therefore $B \cap f(A)$ is empty for any $B \in preO(Y, y_0)$. This is , the mean $f(A) \subset B \subset f(A) \cap Cl_{pre}(B) \neq \emptyset$, this contradicts with the hypothesis which says that the function is $Hpre$ -closed graph . There fore the theorem is true .

Theorem 2.4.7. Let $f: X \to Y$ is $Hpre$ -continuous function and Y is T_2 −space . Then $G(f)$ is strongly $Hpre$ -closed .

Proof.

Suppose that $(x_0, y_0) \in (X \times Y)\backslash G(f)$. Since Y is T_2-space , then there exists a set $B \in O(Y, y_0)$ such that $f(x_0) \notin CL(B)$. But $CL(B)$ is a closed set .Now $Y\backslash CL(B) \in O(Y, f(x_0))$, by Theorem 1.6.5 there exists $A \in HpreO(X, x0)$ such that $f(A) \subseteq Y\backslash CL(B)$. And therefore $f(A) \cap CL(B) = \emptyset$ and so $G(f)$ is strongly $Hpre$-closed .

Theorem 2.4.8. A function $f: X \to Y$ be any surjection and $G(f)$ strongly $Hpre$-closed . Then Y is T_1 .

Proof .

Suppose that c and d be any distinct point of$Y$. Since f is surjection this mean $f(x_1) = d$, for some $x_1 \in X$ and $(x_1, c) \in (X \times Y) \backslash G(f)$. Since the strongly $Hpre$ - closed of $G(f)$ provides $F_1 \in HpreO(X, x_1)$ and $G_1 \in O(Y, c)$ such that $f(F_1) \cap CL(G_1) = \emptyset$. Now, $x_1 \in F_1$, this implies $f(x_1) = d \in f(F_1$. This means that $f(F_1) \cap CL(G_1) = \emptyset$, warranty that $d \notin G_1$. Also from the subjectivity of f we obtain$x_2 \in X$ such that $f(x_2) = c$. Now,$(x_2, d) \in (X \times Y) \backslash G(f)$ and the strongly $Hpre$-closed of $G(f)$ provides $F_2 \in HpreO(X, x_2)$, $G_2 \in O(Y, d)$ such that $f(F_2) \cap CL(G_2) = \emptyset$. Now, $x_2 \in F_2$this implies that $f(x_2) = c \in f(F_2)$, then $c \notin G_2$therefore, we obtain sets $G_1, G_2 \in O(Y)$ such that $c \in G_1$but d $\notin G_1$ while $d \in G_2$but$c \notin G_2$, therefore$Y$ is T_1 space .

Corollary 2.4 9. A function $f: X \to Y$ be any surjection and$G(f)$ strongly $Hpre$ - closed . Then Y is pre - T_1space .

Proof.

A similar of the proof Theorem 2.4.8 .

Corollary 2.4.10.A function $f: X \to Y$ be any surjection and $G(f)$ strongly $Hpre$-closed . Then Y is $Hpre$-T_1 space .

Proof.

Direct from Theorem 2.2.2 (1) and Theorem 2.4.8.

Theorem 2.4.11. A function $f: X \to Y$ be any injective and $G(f)$ strongly *Hpre*-closed . Then X is $Hpre$ - T_1space .

Proof.

Let $a$and b be any distinct point of X . Since f is injective this mean $f(a) \neq f(b)$ when of obtains that $(a, f(b)) \in (X \times Y)\backslash G(f)$. The strongly *Hpre*-closedness of $G(f)$ provides$F_1 \in HpreO(X, a), G_1 \in O(Y, b)$ such that $f(F_1) \cap CL(G_1) = \emptyset$. So $f(b) \notin f(F_1)$this mean$b \notin F_1$. Also $(b, f(a)) \in (X \times Y)\backslash G(f)$ and strongly *Hpre*- closedness of $G(f)$ then$F_2 \in HpreO(X, b), G_2 \in O(Y, f(a))$with $(F_2) \cap CL(G_2) = \emptyset$, warranty that$f(a) \notin f(F_2)$ and therefore $a \notin F_2$. So $F_1$and $F_2 \in HpreO(X)$, such that $a \in F_1$but $b \notin F_1$while $b \in F_2$but $a \notin F_2$. Hence X is *Hpre*-T_1 .

Corollary 2.4.12.A function $f: X \to Y$ be any bijection and $G(f)$ strongly *Hpre*-closed . Then both X and Y are *Hpre*-T_1 .

Proof.

Direct from Corollary 2.4.10 and Theorem 2.4.11 .

Theorem 2.4.13.A function $f: X \to Y$ be any surjection and $G(f)$ strongly $Hpre$-closed . Then Y is T_2 .

Proof.

Suppose that a and $b$be any distinct point of Y .Since f is surjective this mean $f(x_1) = a$, for some $x_1 \in X$ and $(x_1, b) \in (X \times Y)\backslash G(f)$. The strongly $Hpre$- closedness of $G(f)$ provides $A \in HpreO(X, x_1)$ and $B \in O(Y, b)$ such that $f(A) \cap CL(B) = \emptyset$. Now $x_1 \in A$ this implies $f(x_1) = a \in f(A)$. This mean that $f(A) \cap CL(B) = \emptyset$ warrant that, a $\notin(B)$. Then there exists $M \in O(Y, a)$ such that $M \cap B = \emptyset$. So Y is T_2 .

Corollary 2.4.14.A function $f: X \to Y$ be any surjection and $G(f)$ strongly $Hpre$ -closed . Then Y is pre $-T_2$space .

Proof.

A similar of the proof Theorem 2.4.13.

Corollary 2.4.15 .A function $f: X \to Y$ be any surjection and $G(f)$ strongly $Hpre$ - closed . Then Y is $Hpre - T_2$space .

Proof.

Direct from Theorem 2.2.6 (1) and Theorem 2.4.13.

Conclusions

Let us introduce the following conclusions :

Let (X ,T) be a topological space , then

(1) Every pre-continuous function is $Hpre$-continuous function .

(2) Every g-continuous function function is $Hpre$-continuous function .

(3) Every closed graph is $Hpre$-closed graph .

(4) Every strongly pre-closed graph (resp . strongly closed graph is strongly $Hpre$-closed graph .

But the converse is not true for any of the above points .

Future Works

As future works we have the following ideas :

(**1**) We can define *Hpre*-connected space by using *Hpre*-open set and introduce some of the related theorems .

(**2**) Introduced the definition of *Hpre*-compact space via *Hpre*-open set and we make a comparison with him and with the definition *per*-compact in [**8**]. And study some of its characteristics .

References

[1] D. Andrijevic , " open sets ", Mat. Vesn., vol. 38, no. 93 ,(1986), pp. 24-32.

[2] K. Balachandran , p. Sundaram and J. Maki, "On generalized continuous maps in topological spaces ", Men. Fac. Sci. Kochi Univ. Ser. A Math. , vol. 12 ,(1991), pp. 5-13.

[3] N. Bandyopadhyay and p. Bahttacharyya, "Functions with closed graphs" , Bull . Malays. Math .Sci. Soc. ,vo1.2 , no . 28, (2005) , pp.87-93.

[4] M. Caldas and S. Jafari, "On US spaces", Stud. Cercet. Stiint. Ser. Mat. Univ. Bacau. , vol. 14 , (2005), pp. 13-20.

[5] W. Dunham, "A new closure operator for non- topologies ", Bull. Kyungpook Math. J ., vol. 22 ,(1982) , pp. 55-60.

[6] W. Dunham and N. Levine, "Further result on generalized closed sets in topology", Kyungpook Math. J ., vol. 20 , (1980) , pp. 169-175.

[7] R. Engelking, "General topology " ,Heldermannveriag Berlin , 2nd edition , 1989.

[8] T. Hawary , " On generalized -open sets ", Proyecciones Journo of Mathematics , vol. 32 , no. 1 , (2013) , pp. 47-60.

[9] L.L Herrington and p.p.Long , "Characterizations of -closed spaces" , proc. Amer. Math. Soc. , vol. 48 , (1975) , pp. 469-75 .

[10] A. Kar and p. Bhattacharyya, "Some weak separation axioms ", Bull. Calcutta. Math. Soc., vol. 82 , (1990) , pp. 415-422.

[11] N. Levine "Generalized closed sets in topology " Rend . Del. Cire. Mat. di Palermo, vol. 19, no. 1, (1970) ,pp. 89-96.

[12] P.E. Long, "Function with closed graphs" ,Amer Math. Monthly , vol .76 , (1969) , pp. 2-930.

[13] H. Maki ,J. Umehara and T. Noiri , "Every toplogical space is ", Mem . Fac . Sci. Kochi. univ. Math. , vol.17 ,(1996), pp. 33-42.

[14] S. R . Malghan, "Generalized closed maps," J. Karnatak Univ. Sci. , vol . 27.(1982), pp. 82-88.

[15] A. S. Mashour, M. E. A. El-Monsef and S. N El-Deeb, "On continuous and weak pre continuous function ," Proc. Math. Phys . Soc . Egypt , (1982), pp. 47-53.

[16] S. N .Mashhour and R. prasad,"Some new separation axioms" , Ann. Soc . Sci. Bruxelles , vol. 89 , (1975) , pp. 395-402.

[17] H.H. Mustafa and A.AL-SwidiLuay , "Study of -open sets in the topological spaces ", Acceptable for publication in J.I.M.S ,vol.2 , no 33, (2020).

[18] T. Noiri , "On functions with strongly closed graphs" , Acta. Math. Acad. Sei. Hungar , vol.32 , (1978) , pp. 5-373.

[19] T. Noiri , S. Jafari, R. M. Latif and M. Caldas, "Characterization of function with strongly-closed graphs", Sci. stud. Res. Ser. Math. In form, vol. 19, (2009) , pp. 49-58.

[20] T. M. Nour , "Contributions to the theory of bitopological spaces" , Ph. D. Thesis, Delhi Univ. India , 1989.

[21] P. Padma , S. Jackson rani , A. Senbagapriya and B. Ajitha," On closed sets", International Journal of Recent In the study, vol. 3, (2018) , pp. 470-473.

FSC
www.fsc.org
MIX
Papier | Fördert
gute Waldnutzung
FSC® C083411

Kolibri 360 GmbH
Ferdinand-Jühlke-Straße 7
99095 Erfurt, Deutschland
produktsicherheit@kolibri360.de

Druck:
CPI Druckdienstleistungen GmbH
im Auftrag der
Kolibri 360 GmbH
Ferdinand-Jühlke-Str. 7
99095 Erfurt